Copyright © 2017

ILLUSTRATED BY BAKTHI ROSS
(Pictures are illustrated with an optical mouse)

ISBN 978 -1-922220-24-0

Rainbow Sun

By Bakthi Ross

One day Bilby looked at the sun and wondered what colour the sun was.

So he asked the sun.

"Sun!" He called. "You are so bright I cannot look at you. What colour are you?" The sun did not answer.

Bilby thought he might be too far from the sun to hear him.

So Bilby sailed across the ocean.

Walked over land.

Climbed up the highest mountain
in the world to be closer to the
sun. Bilby called "Sun!" You are so
bright I cannot look at you. What
colour are you?"

"Can you guess what colour I am?" replied the sun.

"Are you yellow like the sunflower?" asked Bilby.

"No, but sometimes I look yellow," said the sun.

"You are blue like the sea," said Bilby.

"No, but I make a cloudless sky look blue," said the sun.

"Orange like an orange," said Bilby.

"No, but sometimes I look orange," said the sun.

"Silver like the silvery moon," said Bilby.

"No," said the sun.

"Red like the fire," said Bilby.

"No, but I can look red at sunrise and sunset," said the sun.

"Black like the shadow,"
said Bilby.

"No, but I can look black at solar eclipse, guess again," said the sun.

"White like the cloud," said Bilby.

"No, but I can look white behind the clouds," said the sun.

"Do you have many colours like the rainbow?" asked Bilby.

"Yes! I have all the colours of light. I am the rainbow coloured sun," said the sun.

"I shine all over the world," said the sun.

"Thank you for your bright light," said Bilby.

The End

www.ingramcontent.com/pod-product-compliance
Lightning Source LLC
Chambersburg PA
CBHW052056190326
41519CB00002BA/241